FARMING LEGENDS

Classic Ford Tractors

An Album of Favorite Ford & Fordson Farm Tractors

By Cletus Hohman

Voyageur Press

Copyright © 2004 by Voyageur Press, Inc.

All rights reserved. No part of this work may be reproduced or used in any form by any means—graphic, electronic, or mechanical, including photocopying, recording, taping, or any information storage and retrieval system—without written permission of the publisher.

Edited by Michael Dregni
Designed by Julie Vermeer
Printed in China

04   05   06   07   08   5   4   3   2   1

Library of Congress Cataloging-in-Publication Data

Hohman, Cletus.
  Classic Ford tractors : an album of favorite Ford & Fordson farm tractors / by Cletus Hohman.
     p. cm. — (Farming legends)
  Includes bibliographical references and index.
  ISBN 0-89658-621-9 (hardcover)
  1. Ford tractors—History. 2. Farm tractors—History. I. Title. II. Series.
  TL233.6.F66H64 2004
  629.225'2—dc22
                          2003015347

Distributed in Canada by Raincoast Books, 9050 Shaughnessy Street, Vancouver, B.C. V6P 6E5

Published by Voyageur Press, Inc.
123 North Second Street, P.O. Box 338,
Stillwater, MN 55082 U.S.A.
651-430-2210, fax 651-430-2211
books@voyageurpress.com
www.voyageurpress.com

Educators, fundraisers, premium and gift buyers, publicists, and marketing managers: Looking for creative products and new sales ideas? Voyageur Press books are available at special discounts when purchased in quantities, and special editions can be created to your specifications. For details contact the marketing department at 800-888-9653.

Legal Notice
This is not an official publication of Ford or Fiat. Certain names, model designations, and logo designs are the property of trademark holders. We use them for identification purposes only. Neither the author, photographer, publisher, nor this book are in any way affiliated with Ford or Fiat.

On the frontispiece:
A 1950 Ford Model 8N brochure.

On the title pages:
A 1950 Ford Model 8N. Owner: Chester Todd. (Photograph by Chester Peterson Jr.)

Inset on the title pages:
A Ford Model 8N on Tractor Stilts.

Inset on the contents page:
British farmers gather around an early Fordson, circa 1917. (Glenbow Archives)

Contents

Introduction
Farewell Horses, Hello Tractors — 7

Chapter 1
Henry Ford, Farmer's Son — 9

Chapter 2
Doodlebugs, Thunder Buggies, and Puddle Jumpers — 13

Chapter 3
The Revolutionary Fordson — 19

Chapter 4
The Little Tractor That Could — 31

Chapter 5
The Refined Ford 8N — 51

Chapter 6
Farmyard Hot Rods — 63

Chapter 7
Carrying on the Tradition in England — 73

Chapter 8
The World Tractor — 83

Bibliography — 92

Ford and Fordson Clubs, Magazines, and Resources — 93

Index — 95

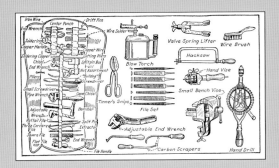

Above: **Recommended Fordson Tool Kit**

Facing page: **1939 Ford-Ferguson Model 9N**
Henry Ford's 9N signaled the second Ford revolution to arrive in the farm field. Owner: Dwight Emstrom. (Photograph by Chester Peterson Jr.)

INTRODUCTION

Farewell Horses, Hello Tractors

The Dawn of a New Era in Farming

Just as Henry Ford gave the world inexpensive personal transportation with his Ford Model T and A automobiles, he provided farmers an inexpensive, lightweight mechanical mule in his pioneering Fordson tractor. The result was nothing less than a revolution in agriculture, and many a farmer bade a sad farewell to Old Dobbin as the horse was traded in on a Fordson.

Yet not only did Henry Ford revolutionize farming once, he did it a second time as well. In 1939, he launched his Ford 9N tractor with dramatic advances in implement control and operator safety created by Irishman Harry Ferguson and Ford's own team of engineers, led by Harold L. Brock.

With his subsequent Model 8N, Ford created what many consider the "perfect" tractor. Some 850,000 Model 8Ns were built—more than any John Deere model—and more than half of them are rumored to still be at work today.

Few people can lay claim to having sparked a revolution, but Henry Ford did it several times over. With his Fordson and Ford tractors, he signaled the dawn of a new era in farming.

CHAPTER 1

Henry Ford, Farmer's Son

Giving a Gift Back to His Past

Above: **Henry Ford and his 1907 Automobile Plow.**

Facing page: **1907 Automobile Plow and 1948 Ford Model 8N**
Henry Ford spent some $600,000 of his own money experimenting with tractors in striving to create the ideal steel horse. This second Automobile Plow was updated with a Ford automobile radiator. It was pictured here in a 1948 Ford promotional photograph alongside the new 8N.

Henry Ford was a son of the family farm. He grew up on his family's spread in Dearborn, Michigan, just outside the budding metropolis of Detroit. When he made the trek to the big city, he left agriculture behind, but he never forgot the drudgery of daily farmwork.

In love with tools and all things mechanical, Ford soon found a job as a "steam engineer" operating a steam traction engine built by G. Westinghouse & Company of Schenectady, New York. Steam power—the harnessed energy created by hot water vapors—revolutionized farming as few things had up until the mid 1800s. In the hands of an able engineer, these machines could power themselves through fields, pulling numerous plow bottoms and expanding a farmer's commercial horizons farther than he or she could ever have dreamed.

Even before Ford began work on the horseless carriage that won him fame in 1908 with the introduction of the Model T, he tried his hand at building a tractor. As early as 1906, he crafted his Automobile Plow and continued to experiment to engineer a viable tractor throughout the 1900s and 1910s. His dream was simple yet grand: A mechanical farm tractor would be his gift back to his past.

Above: 1906 Automobile Plow
Henry Ford's first Automobile Plow of 1906 was propelled by a four-cylinder engine and transmission taken from his elegant Ford Model B touring car. The engine displaced 284 ci (4,652 cc) and created 20 hp, similar in specifications to the subsequent Fordson.

Facing page: **1917 Ford-Farkas Prototype**
In 1917, Ford's chief engineer Eugene Farkas developed this prototype with a unit frame as pioneered by the Wallis Cub tractor. The Ford-Farkas was mounted with a 251-ci (4,111-cc) Hercules engine producing 20 hp, and fitted here for testing with an Oliver plow. With the Ford-Farkas, all the basic design elements were set in place for the Fordson.

Above: **1910s Staude Mak-A-Tractor Advertisement**

Facing page: **1910s Doodlebug**
Ma and Pa pose proudly alongside their homemade tractor. Ford's Model T automobile was tremendously popular among rural folk and had a great emancipating effect on their lives. In need of a mechanical mule to do farmwork, many farmers simply converted their automobiles into makeshift tractors. (Glenbow Archives)

CHAPTER 2

Doodlebugs, Thunder Buggies, and Puddle Jumpers

Home-Brewed Iron Horses

They were known affectionately by a legion of different names on different farms: Doodlebugs, Thunder Buggies, Puddle Jumpers, Jitterbugs, and more.

At the dawn of the twentieth century, lightweight and inexpensive farm tractors were not available to the average farmer, so many looked instead at their old Ford Model T automobile with renewed fondness. They gathered up a welding torch, some scrap metal, and set to work, transforming the skeleton of that ancient Tin Lizzy into a brand-new, home-brewed mechanical mule.

Oftentimes, these homemade farm tractors served their ingenious makers for years, if not decades. Some homegrown tractors were mere stopgap vehicles used until the farmer could afford a Fordson. Others were well-engineered machines that rose above the sum of their disparate parts to become legends in their farm communities, bestowing renown on the farmer who eschewed the factory-built variety of tractor.

Facing page: **1910s Smith Form-A-Tractor Advertisement**
One unsung advantage of converting your car into a tractor was the comfortable seating and top-down motoring—with the convertible roof always at the ready.

Right: **1910s Pullford Tractor Brochure**
By 1915, Ford's Model T automobile cost just $450 while even the least-expensive tractors were double that. With credit usually unavailable in those days for the purchase of a true tractor, many enterprising farmers handy with a wrench built their own home-brewed steel horses from old Model T cars. Quick to jump on the bandwagon, a wide variety of firms sprang up to offer their do-it-yourself tractor kits, as with this brochure from the Pullford Company of Quincy, Illinois.

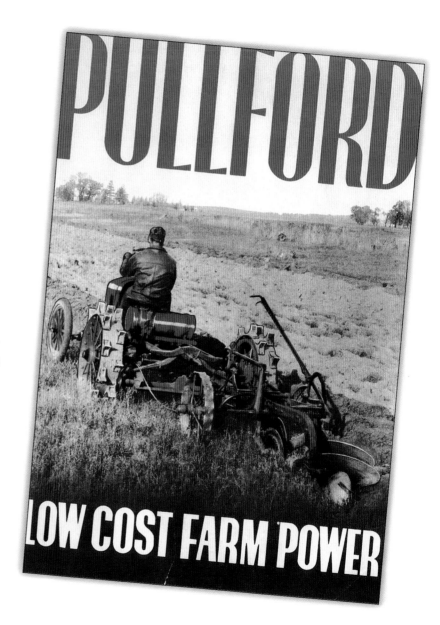

Facing page: **Evidence of Success**
This 1918 brochure for the Staude Mak-A-Tractor featured signed and notarized affidavits from 363 farmers throughout North America lauding the firm's conversion kit that turned your everyday Model T automobile into an iron-willed farm tractor. However, many Ford Model T tractor conversions sadly suffered all the shortcomings of the Model T automobile. The Model T's infamous friction-band planetary transmission was a noted weakness as it was not brawny enough to handle heavy pulling, a basic requirement for plowing. With the debut of Ford's Model A automobile, a better resource for a tractor conversion was happily at hand.

Left: **Thunder Buggy and Happy Family**
Many homemade tractors served their makers for years, if not decades. Along the way, they provided a sign and inspiration to Henry Ford and others that there was a large farm market waiting for lightweight tractors.

CHAPTER 3

The Revolutionary Fordson

A Tractor For Every Farm

Above: **1917 Fordson and Inquisitive Farmers**

Facing page: **1926 Fordson Model F**
When the Fordson was unveiled, the choice for most farmers was not between different tractor makes and models, but between horse or power farming. While many initially held to their horses, in the end most every farmer made the move to the steel horse. Owner: Kukenbecker Tractor Company. (Photograph by Hans Halberstadt)

Steam tractors were monstrous, dangerous, and expensive machines affordable only to the operators of gigantic farms. Family farmers couldn't afford steam power, except at threshing time when a custom threshing crew and its steam traction engine was often hired.

With the refinement of Nikolas Otto's four-cycle, internal-combustion engine, gasoline-fueled tractors were introduced. Soon, garages, workshops, and fledgling factories were building gas tractors of numerous types, from strange Rube Goldberg contraptions to truly innovative machines that took the burden of farming off the farmer and the horse's back and put it on mechanical horsepower.

Henry Ford's Fordson first took the farm by storm in 1918. Suddenly, a trustworthy lightweight tractor was affordable and available for everyone. "Power farming" was here to stay, and by the mid 1920s, 80 percent of the farm tractors at work around the globe were Fordsons.

Above: **Fordson Assembly Line**
Henry Ford became famous not only for his automobiles and tractors but how he built them, perfecting the assembly line that cut manufacturing costs and created uniform products.

Above: **Henry Ford and Fordson**
A farmer's son, Henry Ford was never afraid to get his hands dirty working on his tractor projects.

Facing page: **1910s Ford Tractor Advertisement**
Hearing word that Henry Ford was developing a tractor, a cabal of wily businessmen rushed to launch in 1914 their own Ford Tractor Company in Minneapolis, Minnesota. The firm hoped to cash in on the Ford name—or at least force Henry Ford to buy rights to his own name. But Henry Ford was smarter still. The poor quality of the "Minneapolis" Ford inspired disgruntled Nebraska farmer Wilmot Crozier to run for election to the state legislature, where he penned the famous Nebraska Tractor Test Law of 1919 to prevent farmers from being duped by false tractor promises. And Henry simply called his company Henry Ford & Son and named his machine the Fordson. The "Minneapolis" Ford disappeared as quickly as it arrived, leaving a legacy as history's most infamous tractor.

Ode to the Fordson

The Fordson on the farm arose
Before the dawn at four:
It milked the cows and washed the clothes
And finished every chore.

Then forth it went into the field
Just at the break of day,
It reaped and threshed the golden yield
And hauled it all away.

It plowed the field that afternoon,
And when the job was through
It hummed a pleasant little tune
And churned the butter, too.

For while the farmer, peaceful-eyed,
Read by the tungsten's glow,

THE IMPROVED FORDSON IS A STURDY, RELIABLE, ECONOMICAL AND POWERFUL TRACTOR
FORDSON AGRICULTURAL TRACTOR

Outstanding Improvements

1. More Power
2. Easy Starting
3. New Cooling System
4. New Lubrication System
5. New Ignition System
6. Large Air Washer
7. Redesigned Transmission
8. Longer Wearing Crankshaft
9. Improved Gasoline Carburetor
10. Hot Spot Manifold
11. Heavy Fenders and Platform—Standard Equipment
12. Sixteen Plate Transmission Brake
13. Steel Steering Wheel—Hard Rubber Covered
14. Shock Absorbing Front Coil Spring
15. Automatic Lubrication of Rear Wheel Bearings
16. One Piece Cast Front Wheels
17. Crankcase Ventilation

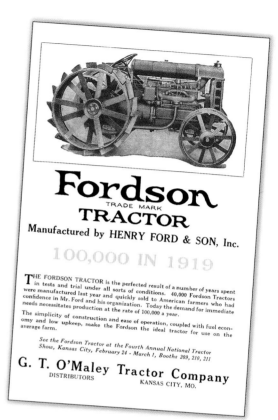

Facing page: **1920s Fordson Brochure**

Above: **1920 Fordson Advertisement**

Left: **1918 Fordson Model F**
The year 1918 marked the first full year of production for Henry Ford's revolutionary lightweight Fordson. The 251-ci (4,111-cc) Hercules four-cylinder L-head engine was rated at 20 hp. This engine was started on gasoline, but ran on kerosene. Owner: Duane Helman. (Photograph © Andrew Morland)

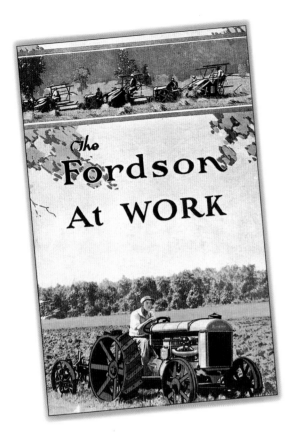

Above: **1920s Fordson Brochure**

Right: **1923 Fordson Calendar**

Facing page: **1926 Fordson Model F**
The Hercules-designed engine was built at Ford from 1921. However, the transmission, gears, and worm differential consumed nearly 50 percent of the 20 hp the engine produced. At the Nebraska test on kerosene, the F recorded only 10.67 drawbar hp. Owner: Fred Bissen. (Photograph © Andrew Morland)

Facing page: **1922 Fordson Model F Hadfield-Penfield Crawler**
The popular Hadfield-Penfield crawler conversion doubled the weight and drawbar pull of the standard Model F. It was manufactured by Hadfield-Penfield Steel Company of Bucyrus, Ohio. Note that the small gasoline tank used for starting can be seen behind the crawler tracks. When hot, the engine was switched to kerosene for operation. Owner: Gene Runkle. (Photograph © Andrew Morland)

Above: **1920s Fordson Trackpull Advertisement**

Right: **Fordson and Happy Farm Family**

Facing page: **Stuck Fordson**
The Fordson wasn't perfect, of course—but then almost any tractor would have got stuck in this muddy field.

Above: **1920s German Fordson Brochure**

Above: **Fordson Rein Controls**
Whoa there! Many farmers were used to driving a horse team and were inexperienced with the feel of a steering wheel. The Rowe Manufacturing Company was just one of several firms offering rein controls for your Fordson so it handled like a horse team.

CHAPTER 4

The Little Tractor That Could

The Ford-Ferguson 9N and 2N

Above: **Humorous Ford Model 9N Postcard**

Facing page: **1942 Ford-Ferguson Model 2N**
When the small, gray 9N was unveiled in 1939, it was not designed to compete with other tractor brands, but with the horse, as engineer Harold L. Brock remembered. In the end, the Ford tractor won out. Owner: Dwight Emstrom. (Photograph by Chester Peterson Jr.)

When Fordson sales faltered during the Great Depression, Henry Ford transferred production to Ireland, then England. It was not until 1937 that Ford again contemplated building an American tractor for American farmers.

Ford had developed several tractor prototypes when he was introduced in 1938 to a visionary Irish engineer named Harry Ferguson who had tractor ideas of his own. Ferguson demonstrated his British-built Ferguson-Brown machine. Ford was impressed and offered to buy the rights, to which Ferguson famously replied, "You haven't got enough money to buy my patents." In the end, the two joined forces, their dealings sealed by just a handshake.

The Ford-Ferguson 9N was launched in 1939 with features created by Ferguson and revised by Ford's engineering team, led by Harold L. Brock. The 9N boasted automatic depth control and a three-point hitch that once again revolutionized farm tractors. It was as hardy as a mule, as strong as an ox, and as trustworthy as a farm dog. The 9N was the little tractor that could.

1937 Ford V-8 Tractor Prototype
Among the many prototypes Ford built before settling on the 9N design was this tractor. It bore obvious resemblance to the Fordson yet was powered by a Ford V-8 engine. With its small chassis and powerful engine, this V-8 tractor would have been a true iron workhorse. Owner: Richard Cummins. (Photographs by Chester Peterson Jr.)

Harry Ferguson and Henry Ford

Ferguson, left, studies the controls

1933 Ferguson "Black Tractor"

Irishman Harry Ferguson was a creative genius who worked on everything from racing motorcycles to airplanes, four-wheel-drive cars to farm tractors. His Ferguson tractor went into production as a joint venture with British gearmaker David Brown. The Ferguson-Brown Model A had Ferguson's revolutionary three-point hydraulic hitch, although it could only be operated on the move.

1939 Ford-Ferguson Model 9N
With its light weight and low price coupled with Ferguson's three-point hitch and automatic draft control, the 9N was just what farmers needed on their small acreages. Henry Ford's first 700 to 800 Model 9N tractors featured cast-aluminum hoods and radiator surrounds with horizontal grille bars. Many collectors today opt to polish the aluminum for its beauty and display the tractor's rarity. Owner: Dwight Emstrom. (Photographs by Chester Peterson Jr.)

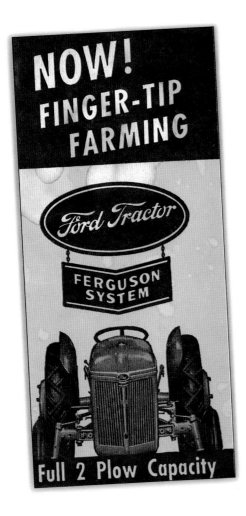

Above: **1940s Ford-Ferguson Farm Ledger**

Right: **Ford-Ferguson Radiator Badges**

Facing page: **Ford-Ferguson 9N and Proud Farmer**

Facing page: **Ford Tractor Sign**
Well-known Ford expert Palmer Fossum of Northfield, Minnesota, stands outside his workshop with an unique porcelain Ford sign. (Photograph by Chester Peterson Jr.)

Above: **1939 Ford-Ferguson Model 9N**
The Ford-Ferguson was christened the 9N with "9" signifying the year of introduction, 1939. Owner: Dwight Emstrom. (Photograph by Chester Peterson Jr.)

Top: **Ferguson System Sign**

Bottom: **1939 Ford-Ferguson Model 9N**
Close-up view of the famed three-point hitch. (Photograph by Hans Halberstadt)

Left: **1940 Ford-Ferguson 9N**
Both together and separately, Ford and Ferguson offered a wide catalog of optional equipment and accessories for their tractors, including this rare and handsome cab. Realizing the popularity of the Ford-Ferguson machines, numerous aftermarket suppliers also provided a variety of components, much as they did earlier with the Fordson. Owner: Ron Stauffer. (Photograph by Chester Peterson Jr.)

Above: **1939 Ford-Ferguson Model 9N**
The 9N's four-cylinder side-valve powerplant was developed from half of a Ford V-8 truck engine. It displaced 119.6 ci (1,961 cc) and produced 23 PTO hp via a three-speed gearbox. Ford also used other car and truck parts in the tractor where possible. Owner: Dwight Emstrom. (Photograph by Chester Peterson Jr.)

1939 Ford-Ferguson Model 9N

This 9N, serial number 266, was the first Ford-Ferguson shipped to California. The 9N was launched in 1939 and continued in production until 1942, when it was superseded by the Model 2N. Owner: Kukenbecker Tractor Company. (Photograph by Hans Halberstadt)

Facing page: **1941 Ford-Ferguson 9N**
A hard-working and well-used 9N. (Photograph by Chester Peterson Jr.)

Left: **Ford-Ferguson Accessories Catalog**

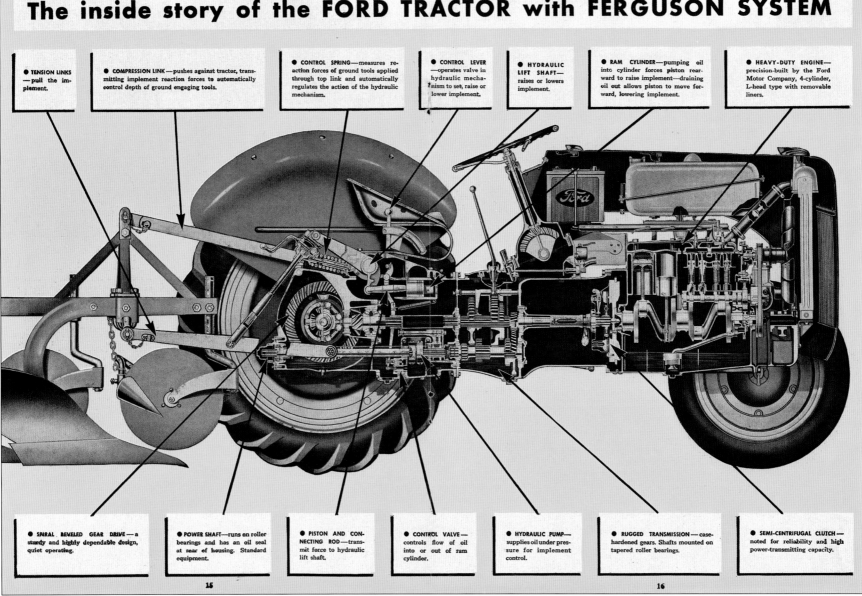

Facing page: **1940s Ford-Ferguson Brochure**

Above: **Ferguson System Dealer Clock**

Right: **1942 Ford-Ferguson Brochure**

Left: **1945 Ford-Ferguson Model 2N**
Like the 9N, the 2N got its name from its year of introduction, 1942. The 2N retained 9N's side-valve four-cylinder engine, displacing 119.6 ci (1,961 cc) and producing 23 PTO hp via a three-speed gearbox. Owner: Palmer Fossum. (Photograph by Chester Peterson Jr.)

Below: **1945 Ford-Ferguson Model 2N**
This 2N rides atop optional sand wheels designed to "float" the 2,400-lb (1,080-kg) tractor through sand and sandy soils. Owner: Palmer Fossum. (Photograph by Chester Peterson Jr.)

Facing page: **1945 Ford-Ferguson Model 2N**
An original weather cover draped over a 2N. Owner: Palmer Fossum. (Photograph by Chester Peterson Jr.)

1940s Ford-Ferguson Brochure

Above: **1942 Ford-Ferguson Model 2N**
Like the previous 9NAN model, a 2NAN was available with an engine modified to run on distillate fuel. This 2N was powered by gasoline. Dwight Emstrom. (Photograph by Chester Peterson Jr.)

Facing page: **1940s Ford-Ferguson B-NO-40 Moto-Tug**
A small production run of aircraft tow tractors were built for the U.S. military by Ford from converted 2N farm tractors. Owner: Dwight Emstrom. (Photograph by Chester Peterson Jr.)

CHAPTER 5

The Refined Ford 8N

A Good Things Gets Better

Above: **1940s French Ford Model 8N Brochure**

Facing page: **1951 Ford Model 8N High-Crop**
After Henry Ford II took over the Ford Motor Company from his father in 1945, he dissolved the Handshake Agreement with Harry Ferguson. In mid 1947, Ford then launched its own Model 8N tractor without the Ferguson name. Owner: Dwight Emstrom. (Photograph by Chester Peterson Jr.)

Given their brilliant yet fiery personalities, Henry Ford and Harry Ferguson's Handshake Agreement was almost inevitably doomed from the start. When Ford died in 1947, company control was relegated to his son, Henry Ford II. Soon after, the pact with Ferguson was dissolved, resulting in a rancorous, expensive lawsuit that plowed on for years.

While their lawyers battled it out in court, Ford launched a refined version of the 9N and 2N as the 8N. Not to be outdone, Ferguson also debuted his own Ferguson TE-20 based on the old 9N. The paint schemes and radiator badges might have been different, but underneath they were similar machines based on a shared lineage.

During its production run, some 850,000 8N tractors were built, and more than half of them are rumored to still be at work five decades later. The 8N was, in many farmers' minds, the "perfect" tractor.

1951 Ford Model 8N High-Crop

Ford's 8N was a refined version of the 9N and 2N with numerous mechanical upgrades. The engine still displaced 119.7 ci (1,961 cc) but now created 26 PTO hp, a power increase of more than 10 percent. The old three-speed gearbox was replaced by a more-flexible four-speed. With an optional auxiliary, twelve forward speeds could be had. Owner: Dwight Emstrom. (Photographs by Chester Peterson Jr.)

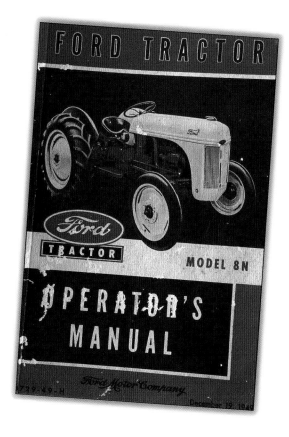

Above: **1940s Ford Model 8N Operator's Manual**

Left: **1947 Ford Model 8N Advertisement**

Ford Tractors Sign

Facing page: **1950s Ferguson**
After parting with Ford, Harry Ferguson quickly regrouped and launched his own tractor, the TE-20 of 1946, beating Ford to the draw by almost two years. Ferguson's machine was obviously based on the Ford-Ferguson 9N and 2N—right down to the gray paint scheme. Built in Detroit, Michigan, and in Coventry, England, the trustworthy "Fergie" won many fans. The TE-20 was available with a variety of fuel options and later evolved into the TO-30 and TO-35 before Ferguson, tired of the tractor field, merged his company with Massey-Harris in 1953. (Photograph by Hans Halberstadt)

Above: **1952 Ford Model 8N**
As with the 9N and 2N, the 8N's name signified its year of introduction, 1948—even though the new tractor was unofficially available in late 1947. Owner: Kukenbecker Tractor Company. (Photograph by Hans Halberstadt)

1940s Ford Model 8N State Fair Display

Above: **1952 Ford Model 8N High-Crop**
The most obvious change introduced by the 8N was the new paint scheme. Gone was the old gray drab of the 9N and 2N, replaced by a light gray and red paint job that made the tractor look thoroughly modern. (Photograph by Chester Peterson Jr.)

Right: **1950 Ford Model 8N**
The 8N also received an improved hydraulic lift, steering mechanism, and braking system. All in all, it was a dramatically refined tractor, making the old 9N appear almost crude in comparison. Owner: Chester Todd. (Photograph by Chester Peterson Jr.)

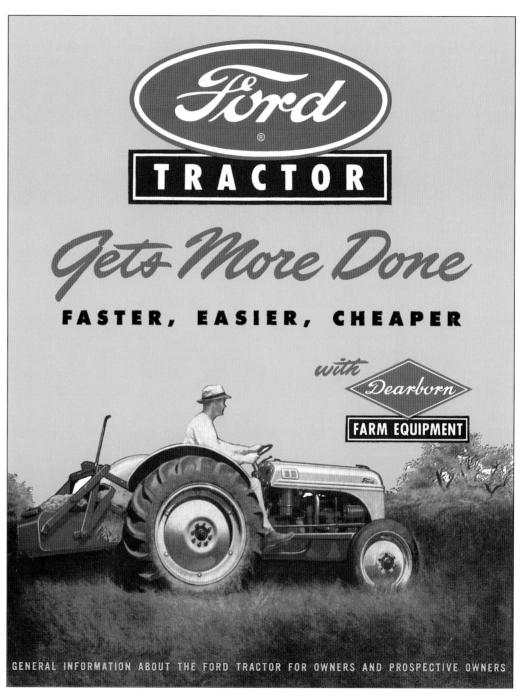

Above: **1940s Ford Model 8N**

Left: **1940s Ford Model 8N Brochure**

1952 Ford Model 8N
Just as many early farmers used their Ford Model T and A automobiles to do everything from run clothes-washing machines to wood saws, equipment was available to turn the 8N to any farm chore. This tractor was fitted with a rear-mounted rotary saw. Owner: Dwight Emstrom. (Photograph by Chester Peterson Jr.)

1940s Ford Model 8N
As with the Fordson and Ford-Ferguson 9N and 2N, numerous aftermarket suppliers rushed to offer components to augment the 8N. This 8N was equipped with Bombardier "half-track" crawler tread. Owner: Palmer Fossum. (Photograph by Chester Peterson Jr.)

Above: **1950 Ford Model 8N**
After its rancorous parting with Ferguson, Ford set up its own implement firm, Dearborn Farm Equipment, to supply Ford tractor owners with all manner of tillage and harvesting equipment. Dearborn also offered other options, such as the grader mounted to the front of this 8N. Owner: Dwight Emstrom. (Photographs by Chester Peterson Jr.)

Right: **1940s Ford Model 8N**
Essential equipment for farmers in snow country, a Dearborn bulldozer blade was mounted to the front end of this Minnesota 8N. Owner: Palmer Fossum. (Photograph by Chester Peterson Jr.)

Above: **1950s Funk Aircraft Company Advertisement**

Facing page: **Ford-Funk Model 8N V-8**
The Funk brothers' conversion was a true hot rod in the best automotive sense, shoehorning a big engine into a small chassis to create a high-powered machine. Owner: Palmer Fossum. (Photograph by Chester Peterson Jr.)

CHAPTER 6

Farmyard Hot Rods

The Ford-Funk Conversions

The Ford-Funk tractors were neither dragstrip racers nor main-street hot rods. Yet by shoehorning a Ford six-cylinder or V-8 truck engine into an N Series tractor, farmers suddenly had a machine with all the muscle, torque, and pulling power they could ever need.

Twins Joe and Howard Funk began by building lightweight airplanes using Ford car engines at their base in Coffeyville, Kansas. In 1948, Joe happened to visit a Ford dealership where the owner had built his own hot-rodded 9N. It sparked an idea, and the Funk Aircraft Company was soon offering conversion kits to farmers in need of more brawn in their tractors.

Over the years, many 8Ns were hopped up with larger engines thanks to the resources of the Funk kits. Today, these farmyard hot rods are among the rarest and most collectible of all Ford tractors.

Facing page: **1950s Ford V-8 Prototype**
In the late 1940s, Ford experimented with a more-powerful version of its Model 8N mounted with a Ford V-8 engine. The prototype never went into production, and farmers in search of more muscle had to buy a Funk kit or build their own hot rod.

Right: **Ford V-8 Radiator Badge**

1950s Ford-Funk Model 8N V-8
Ford V-8 power in the small 8N chassis created a lightweight tractor with all the brawn a farmer could ever wish for. The sole concern was at times having too much muscle under the hood and spinning the rear wheels to sink the tractor into a field's soil. Owner: Palmer Fossum. (Photograph by Chester Peterson Jr.)

1950 Ford-Funk Model 8N V-8

Most farmers who converted their 8Ns to V-8 power opted for the Ford 8BA flathead engine. Displacing 239 ci (3,915 cc), the V-8 produced a whopping 85 hp. More importantly for most farm users, the engine created a stump-pulling 187 ft-lb of torque. Owner: Delbert Heusinkveld. (Photographs by Chester Peterson Jr.)

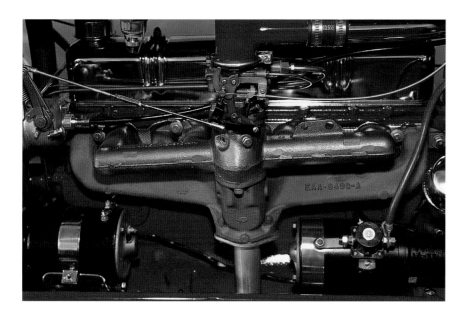

Facing page: **1950s Ford-Funk Model 8N V-8**
Ford V-8 power looked and sounded best when channeled through open pipes. Owner: Palmer Fossum. (Photograph by Chester Peterson Jr.)

Left top and bottom: **1950s Ford-Funk Model 8N Six-Cylinder**
Ford's overhead-valve six-cylinder engine displaced 225 ci (3,686 cc) and provided an 8N with 80 hp and 182 ft-lb of torque. Owner: Johnny Grist. (Photograph by Chester Peterson Jr.)

Above: **1950 Ford-Funk Model 8N V-8**
An estimated 5,000 Ford 8N tractors were converted to six-cylinder or V-8 power with Funk kits—a small figure compared to the estimated 850,000 Model 8N tractors produced. Owner: Ron Stauffer. (Photograph by Chester Peterson Jr.)

CHAPTER 7

Carrying on the Tradition in England

British Fordson Tractors

In 1928, Henry Ford founded subsidiaries in England and Ireland to sell his Fordson machines. With the decline of sales in the United States during the Great Depression, tractor production was moved in 1929 to Cork, Ireland, then in 1933 to Dagenham, England.

The British Fordson soldiered on, being exported around the globe and even in limited numbers back to North America. Revisions and refinements were made over the years, but the basic tractor was the same stalwart machine launched in 1918.

It was not until the end of World War II that a major new model was unveiled, titled fittingly the Fordson Major. The updated and uprated E27N Major was soon followed by the E1A New Major in 1952 with diesel or TVO (Tractor Vaporizing Oil) fuel options. Super Major, Power Major, Dexta, and Super Dexta versions kept the English line alive until the next generation of Ford tractors was unveiled.

Above: **1930s Fordson Major Advertisement**

Facing page: **1941 Fordson Model N**
After the start of World War II in 1939, the Fordson paint scheme was changed from orange to green to reduce the visibility of the Fordson as a target for enemy aircraft. This Model N pulls a 1937 horse-drawn Somerset root drill. (Photograph © Andrew Morland)

FARM WITH A NEW FORDSON

FORDSON

Ford Motor Co., Dearborn, Mich.

Air Cleaner: Handy, wet.
Carburetor: Stromberg or Kingston, 1¼ in.
Clutch: Own, disc.
Governor: Handy, flyball.
Ignition: Robert Bosch, high tension magneto.
Lighting: Own, electric, optional.
Oil Filter: ----------------------
Radiator: Own, fin tube.
Radiator Cover: ----------------------
Spark Plugs: Champion 1 Com.
Starting: ----------------------
Data: H.P.—Neb. Test No. 173 (kerosene): Max. Belt 23.24; Max. D.B. 13.60; Max. lbs. pull 3,289 at 1.55 m.p.h. Weight as tested (with operator) 3,820 lbs. Test No. 174 (gasoline): Max. Belt 29.09; Max. D.B. 18.30; Max. lbs. pull 3,202 at 1.89 m.p.h. Number of plows recommended: Two, 14 in.
Engine: Own "N"; 4⅛x5, 1,100 r.p.m., 4 cylinders, vertical, L-head, cast en bloc.
Pulley: 9½ x 6½, 1,100 r.p.m. and 2.730 f.p.m. at normal engine speed.
Speeds: M.P.H. forward 2.2 to 4.3, and 1.67 reverse.

THERE are many new features in the new Fordson agricultural tractor. These include new high-carbon steel crankshaft. Improved heat treatment of gear steel to increase strength and surface hardness. New fenders and shields to protect driver from dust. And a *special* feature that gives you a choice of three fuel systems—gasoline carburetor, kerosene vaporizer, or fuel oil vaporizer.

Write for full details concerning this sturdy, economical Fordson. This tractor that is built to give years of faithful service. Coupon below is for your convenience.

NOTE: *Fordson parts are always available for all models of Fordsons through your nearest Ford or Fordson dealer. Insist on genuine Fordson parts.*

SHERMAN & SHEPPARD, Inc.

2nd Ave. & 34th St., Brooklyn, N. Y.　　　Phone SUnset 6-3360

Facing page: **1930s American Fordson Advertisement**

Left: **1937 Fordson Model N**
Front end of a Dagenham-built Model N, showing the crank starter. (Photograph by Hans Halberstadt)

Above: **1935 Fordson Model N**
Around 200,000 Fordsons were manufactured at Dagenham, England, from 1933 through 1945. Owner: John Carwood. (Photograph © Andrew Morland)

The FORDSON TRACTOR does

- Under average soil conditions, the Fordson Tractor will plough eight to fourteen acres a day, with two or three 14″ bottoms. That's *better* than a good day's work! And when you figure how little the Fordson costs to run, and it takes the meanest, toughest farm jobs in its stride—you'll agree that this tractor that has satisfied more than three-quarters of a million farmers, may very well satisfy you!

The Fordson really does the work of 6 or 8 horses. It's always ready to work — day or night. It can pull a 6′ one-way disc, or a 7′ to 10′ double disc harrow, or operate a 24″ to 26″ thresher with all attachments. The Fordson *makes* mo by doing your work easily and quickly. It has all the power a dependability that you've associated for years with the Ford na And the Fordson *saves* money because (1) it costs less to buy t

Shown above is the Fordson agricultural tractor with 11.25 x 24 pneumatic rear tires and 6.00 x 16 fronts on spoke-type wheels. Weights are used to increase traction. Water ballast may be used in the tires.

Cross sectional view of the Agricultural Tractor with spade lugs is steel rear wheels. (3) The truck-type steeri

FORDSON AGRICULTURAL TRACTOR SPECIFICATIONS

ENGINE
Four cylinder, four-stroke, cylinders cast en bloc. Cylinder bore 4⅛″. Piston stroke 5″. Piston displacement 267 cu. in. Firing order 1, 2, 4, 3. Statically and dynamically balanced crankshaft carried on 3 main bearings, 2″ diameter by 3″ long. Connecting rod bearings 2″ diameter by 2¼″ long. Cast-iron pistons, with 3 compression and 1 oil control piston rings fitted above the piston pin. Side valves of special steel alloy. High compression cylinder head for running on gasoline. Low compression cylinder head for running on kerosene or fuel oil.

LUBRICATION
Splash system with oil circulation maintained by oil thrown off flywheel by centrifugal action. Ducts lead oil to main bearings and timing gears. Crankcase ventilation. Oil capacity 2¼ imperial gallons.

COOLING SYSTEM
Thermo-syphon, impeller assisted. Efficient cooling ensured by vertical tube type radiator with very large reinforced tanks, in conjunction with ample water jackets around cylinder block and cylinder head, and four-blade fan which draws 1,700 cu. ft. of air per minute through the radiator. Cylinder water inlet is 2¾″ diameter, outlet 2⅝″ diameter. Water capacity, 10 imperial gallons.

IGNITION SYSTEM
High-tension magneto with impulse coupling for easy starting, driven by a helical gear from camshaft gear. Manual advance and retard controlled by lever on dash.

GOVERNOR
Centrifugal type, adjustable to maintain any desired engine speed. Controlled from dash.

FUEL SYSTEM
Gravity feed from 17½ imperial gallon overhead tank, through a sediment bulb which filters all fuel before it reaches the vapourizer or carburetor. All tractors are equipped with double fuel tank. When operating on kerosene or fuel oil a one-gallon auxiliary compartment inside the main tank is used for gasoline.

VAPOURIZERS
Tractor fitted with low compression cylinder head for running on kerosene or fuel oil has special vapourizer, according to type of fuel used. Each of these two special types of vapourizers has an adjustable mixture control and exhaust-heated plate to ensure economy an proper vapourization of fuel used.

CARBURETOR
Tractor fitted with high compression cylinder head fo running on gasoline has special down-draft carbureto

AIR CLEANER
A primary centrifugal air cleaner, at top of air intak pipe, draws air to the oil bath air cleaner. This ensure that all air used by the engine is thoroughly cleane minimizing wear on all engine parts. Capacity of o bath air cleaner is 6 pints.

STEERING SYSTEM
Worm and sector. Ratio 17 to 1. All steering arm and rods are of heavy construction, for long life. Easil replaceable steel bushings are fitted to steering sha and spindle bodies.

FRONT AXLE
Heavy drop forging, heat treated to provide maximu strength. It is mounted at the centre of the front o the engine on a rubber buffer and a plunger which tak up all road shocks. Heavy radius rods are fitted to take up thrust and maintain alignment.

work of 6 or 8 horses

tractor of comparable power; (2) optional equipment is low cost; (3) it's economical to run; (4) simple design and rugged construction make for long, trouble-free operation; (5) replacement ...s and repairs, when you eventually need them, cost little.

In addition to the two Fordson models shown here, there is a Row-Crop Fordson with 50" skeleton type rear wheels and 4" overhanging lugs, which has a high clearance and adjustable rear wheel tread, for all planting and cultivating uses. Your Ford Dealer strongly invites you to make your own test of Fordson power and economy on your farm. There is no cost or obligation for this!

Ford Motor Company of Canada, Limited, whose policy is one of continuous improvement, reserves the right to change specifications and prices at any time without notice or incurring liability to purchasers.

(1) The transmission with optional gear ratios. (2) The 9-inch ...th air cleaner. (5) The radiator shutter.

And here is the Fordson agricultural tractor equipped with 42" 9-inch rim rear wheels with 4½" forged offset lugs. More than three-quarters of a million Fordson tractors work the world's farms ... cut farmers' labour ... increase their earnings.

story of rugged power, long life, economy!

...RANSMISSION
... or 7.75 optional both with standard or special ratio ...ars. Constant mesh sliding selective type with three ...eeds forward and reverse. All shafts run on ball or ...ller bearings. Clutch of multiple disc type. Oil ...pacity 3¾ imperial gallons.

...RAKE
...ultiple disc on transmission operated by clutch pedal. ...w-Crop Tractor has automatic turning brakes, en-...ling tractor to pivot on either rear wheel.

...EAR AXLE
...mi-floating, four pinion differential, mounted on roller ...arings. Worm drive, phosphor bronze worm wheel, ... axle shaft, 17 to 1 ratio.

...HEELS
...gricultural—Spade lugs. Front: 30" x 5" heavy cast ...on, mounted on adjustable roller bearings. Rear: 42" ...ameter, 9" wide, equipped with 4½" forged offset lugs. ...gricultural—Pneumatic Tires. Front 4.50 x 16 wheels ...th weights and 6.00 x 16 tires. Rear: 8.00 x 24 wheels ...th inner and outer weights, 11.25" x 24" ground ...ip tires. Row-Crop—Spade Lugs. Front: 24" x 4" with

1½" skid rings. Rear: 50" skeleton steel wheels with 4" overhanging lugs.

TOOLS
Full equipment of tools. Tool box mounted on dash. All pneumatic tire equipped tractors are supplied additionally with engine-operated tire pump, jack and two tire irons.

DRAW-BAR
4 Wheel Tractor has stationary type with 7" lateral adjustment; height of draw-bar from ground, 13½". Row-Crop tractor has swinging type.

PULLEY WITH CLUTCH (At extra charge)
Width 6½", diameter 9½" spiral bevel gears. Speed 1100 r.p.m. Belt speed 2728 ft. per minute.

POWER TAKE-OFF (At extra charge)
Separate from pulley. Operates at speed of 534 r.p.m. at engine speed of 1100 r.p.m. This is only supplied with 4.3 transmissions. Combined pulley and power take-off can be supplied for use with 7.75 transmissions.

WEIGHT
Tractor, with driver and fuel, 3,550 lbs. when mounted on steel, 3,950 lbs. on rubber. Row-Crop 3,800 lbs.

DIMENSIONS
4 Wheel Agricultural tractor—spade lugs. Wheelbase, 63". Overall length of tractor (with fenders) 106 7/8"; overall width 62⅛"; overall height 55 7/8". Ground clearance, 9". Turning circle 23'6".

STANDARD EQUIPMENT
High Compression Head (Gasoline)
Gasoline Carburetor
Double Compartment Tank
Vertical Exhaust Pipe
Straight Front Axle
Wheels—Spade lugs and Fenders
4.3 (top gear) Transmission—Standard Ratio.

OPTIONAL EQUIPMENT (At no extra charge)
Low Compression Head (Kerosene or Fuel Oil)
Kerosene Vapourizer or Fuel Oil Vapourizer
Rear (horizontal) Exhaust Pipe
4.3 (top gear) Transmission—Special Ratio
7.75 (top gear) Transmission—Standard Ratio
7.75 (top gear) Transmission—Special Ratio.

OPTIONAL EQUIPMENT AT EXTRA COST
Power Take-Off   Pulley with Clutch
Electric Lighting Equipment.

Above: **1937 Fordson Model N** After production of the Fordson ceased in the United States, the tractors continued to be built in Ford's Cork, Ireland, plant before eventually being transferred to Dagenham, England, where this Model N was produced. Owner: Kukenbecker Tractor Company. (Photograph by Hans Halberstadt)

Left: **1930s Canadian Fordson Brochure**

Facing page: **1946 Fordson Major E27N**
The Model N's four-cylinder L-head engine was designed back in 1917. Powered by the upgraded Model N engine, the E27N was marketed as a 27-hp tractor but was actually rated at 19.1 drawbar and 28.5 belt hp. At the peak of production in 1948, more than 50,000 were produced. Owner: Ivan Sparks. (Photograph © Andrew Morland)

Right: **1940s Australian Fordson Major Advertisement**

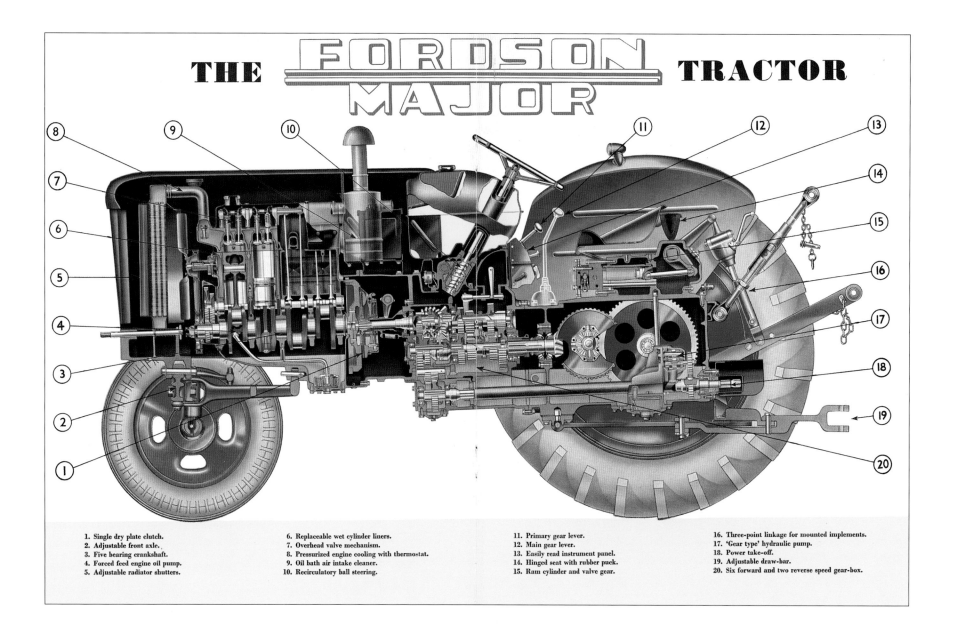

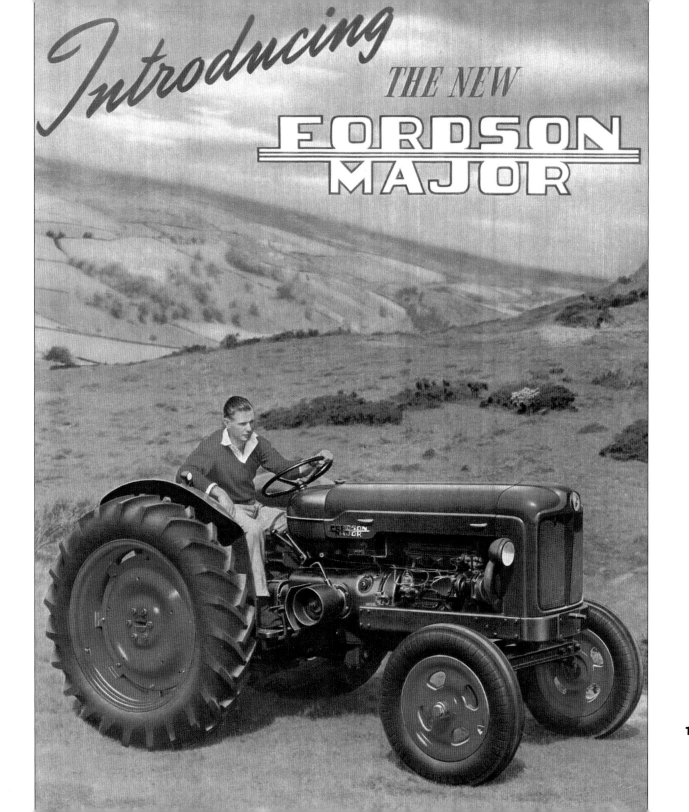

1950s Fordson Major Brochures

CHAPTER 8

The World Tractor

From the N Series to a Worldwide Model Line

Above: **1953 Ford Model NAA Advertisement**

Facing page: **1964 Ford Model 2000 Diesel**
Ford's new World Tractor line was launched in 1961 with the Model 6000 and expanded in 1962 with the 5000 and 2000 machines. It was the start of a bold new era at Ford. Owner: McGinn Sales. (Photograph by Chester Peterson Jr.)

Ford always offered just one tractor model, albeit with a wealth of options and specialized variations. Yet beginning in 1955, the line proliferated with a full spectrum of tractor choices.

In 1953, Ford launched its Model NAA Golden Jubilee tractor to celebrate the firm's fiftieth anniversary. The NAA was followed in the rest of the 1950s by the 600, 700, 800, and 900 Series with model lineups offering a variety of engine options, transmissions, hydraulics, and other equipment.

It was in 1961, however, that Ford truly began looking at the tractor market with a global vision. Ford Tractor Operations (soon retitled the Ford Tractor Division) was sparked, and the new Thousand Series made its debut. Henry Ford's steel horse had truly become a World Tractor.

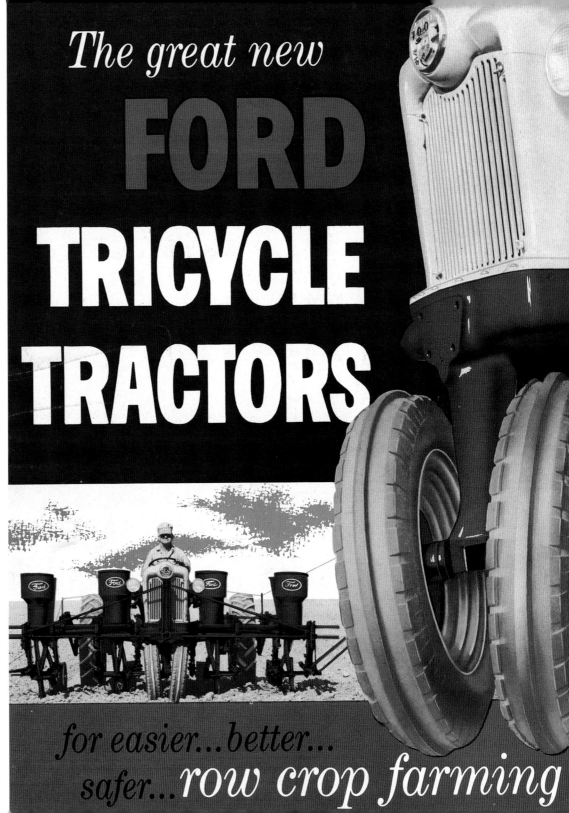

Above: **1953 Ford Model NAA Advertisement**
Ford replaced the 8N with its new Model NAA in 1953 to celebrate the firm's fiftieth anniversary. Known as the Golden Jubilee model, the NAA boasted a new overhead-valve Red Tiger engine displacing 134 ci (2,195 cc) and producing 30 hp. A refined hydraulic system, better governor, and a host of other lesser improvements also marked the new tractor.

Right: **1961 Ford Tractor Brochure**
Ford's lineup of tractor models proliferated by 1960 to include the Fordson Major Diesel from England and the American-built 600, 700, 800, and 900 Series. From two- to four-plow power, Ford had it all.

Above: **Ford Tractors Neon Dealer Sign**

Right: **1950s Ford Model 601 Workmaster**
Starting in 1955, Ford broke away from its tradition of offering just one tractor model and unveiled a whole line. The 600 Series of 1955–1957 was followed by the 1958–1961 601 Series. Owner: Kukenbecker Tractor Company. (Photograph by Hans Halberstadt)

Left: **Ford Model 961 Diesel Advertisement**

Above and facing page: **1960 Ford Model 971 LPG**

The 971 was at the top of Ford's lineup at the dawn of the 1960s. Powered by a 172-ci (2,817-cc) overhead-valve four-cylinder engine, it had 50 hp available through ten speeds via the Select-O-Speed powershift transmission. The "7" in the model name denoted the Select-O-Shift, hydraulic lift, and a live PTO. Owner: Dwight Emstrom. (Photograph by Chester Peterson Jr.)

1964 Ford Model 4000

Ford's 801 became the 4000 with a new paint scheme and new sheet metal as part of the World Tractor line. Owner: Paul Martin. (Photograph by Chester Peterson Jr.)

Top: **1960s Ford Model 4000 Assembly Line**

Above and facing page: **1967 Ford Model 6000 LPG**
The 6000 was launched in 1961, the first of Ford's World Tractors. A large tractor at 10,000 lb (4,500 kg), the 6000 needed every drop of the 60 drawbar hp its engine fathered. As part of the new World Tractor concept, it was available in a wide variety of row-crop versions with either a wide or tricycle front end and fueled by either gasoline, diesel, or LPG. Owner: Dwight Emstrom. (Photograph by Chester Peterson Jr.)

Bibliography

Baldwin, Nick and Andrew Morland. *Classic Tractors A-to-Z*. Stillwater, Minnesota: Voyageur Press, 1998.

Cherouvrier, Jean and Jean Noulin. *Tracteurs Ferguson: Un Homme, un Systeme, un Mythe*. Paris: ETAI, 1999.

Dregni, Michael, ed. *This Old Tractor: A Treasury of Vintage Tractors and Family Farm Memories*. Stillwater, Minnesota: Voyageur Press, 1998.

Hohman, Cletus. *Classic Farm Tractors: An Album of Favorite Farm Tractors from 1900–1970*. Stillwater, Minnesota: Voyageur Press, 2001.

Pripps, Robert N. and Andrew Morland. *The Field Guide to Vintage Farm Tractors*. Stillwater, Minnesota: Voyageur Press, 1999.

Pripps, Robert N. and Andrew Morland. *Vintage Ford Tractors: The Ultimate Tribute to Ford, Fordson, Ferguson, and New Holland Tractors*. Stillwater, Minnesota: Voyageur Press, 1997.

1964 Ford Model 4000
Ford's 801 became the 4000 with a new paint scheme and new sheet metal as part of the World Tractor line. Owner: Paul Martin. (Photograph by Chester Peterson Jr.)

Ford and Fordson Clubs, Magazines, and Resources

Clubs and Magazines

9N–2N–8N–NAA Newsletter
Gerard and Bob Rinaldi
P. O. Box 275
East Corinth, VT 05040–0275

Ford/Fordson Collectors
Association
645 Loveland-Miamiville Road
Loveland, OH 45140

General Parts

Ag Tractor Supply
Box 276
Stuart, IA 50250

All Parts International, Inc. (API)
3215 West Main Avenue
Fargo, ND 58103
www.stpc.com

Bob Martin Antique Tractor Parts
5 Ogle Industrial Drive
Vevay, IN 47043
www.venus.net~martin

The Brillman Company
Box 333
Tatamy, PA 18085
www.brillman.com

Central Michigan Tractor & Parts
2713 N. U.S. 27
St. Johns, MI 48879

Central Plains Tractor Parts
712 North Main Avenue
Sioux Falls, SD 57102

Colfax Tractor Parts
Rt. 1, Box 119
Colfax, IA 50054

Dengler Tractor
6687 Shurz Road
Middletown, OH

Detwilier Sales
S3266 Highway 13 S.
Spenser, WI 54479
715-659-4252

Discount Tractor Supply
Box 265
Franklin Grove, IL 61031

Draper Tractor Parts, Inc.
Rt. 1, Box 41
Garvield, WA 99130

Fresno Tractor Parts
3444 West Whitesbridge Road
Fresno, CA 93706

Iowa Falls Tractor Parts
Rt. 3, Box 330A
Iowa Falls, IA 50126

Restoration Supply Co.
Dept. AP96 Mendon Street
Hopedale, MA 01747

South-Central Tractor Parts
Rt. 1, Box 1
Leland, MS 38756

Southeast Tractor Parts
Rt. 2, Box 565
Jefferson, SC 29718

Steiner Tractor Parts, Inc.
G-10096 S. Saginaw Road
Holly, MI 48442
www.steinertractor.com

Surplus Tractor Parts Corp.
Box 2125
Fargo, ND 58107

Yesterday's Tractors
P. O. Box 160
Chicacum, WA 98325
www.ytmag.com

Specialized Parts

A-1 Leather Cup and Gasket
Company
2103 Brennan Circle
Fort Worth, TX

Agri-Services
13899 North Road
Alden, NY 14004
Specializing in wiring harnesses

Antique Gauges, Inc.
12287 Old Skipton Road
Cordova, MD 21625
Specializing in gauges

Jorde's Decals
935 Ninth Avenue N.E.
Rochester, MN 55906
www.jordedecals.com

Lubbock Gasket & Supply
402 19th Street, Dept. AP
Lubbock, TX 79401

Jack Maple
Rt. 1, Box 154
Rushville, IN 46173
Decals for a wide variety of applications and models

M. E. Miller Tire Co.
17386 State Highway 2
Wauseon, OH 43567
www.millertire.com

Nielsen Spoke Wheel Repair
Herb Nielsen
3921 230th Street
Esterville, IA 51334

Olson's Gaskets
3059 Opdal Road E.
Port Orchard, WA 98366
www.olsonsgaskets.com

Omaha Avenue Radiator Service
100 E. Omaha Avenue
Norfolk, NE 68701

Tractor Steering Wheel Recovering
and Repair
1400 121st Street W.
Rosemount, MN 55068

Tractor Manuals

Clarence L. Goodburn
Literature Sales
101 W. Main
Madelia, MN 56062

Intertec Publishing
P. O. Box 12901
Overland Park, KS 66282
www.intertecbooks.com

Jensales Inc.
P. O. Box 277
Clarks Grove, MN 56016
www.jensales.com

King's Books
P. O. Box 86
Radnor, OH 43066

Yesterday's Tractors
P. O. Box 160
Chicacum, WA 98325
www.ytmag.com

Index

Brock, Harold L., 7, 31
Brown, David, 34
Crozier, Wilmot, 20
Dearborn Farm Equipment, 60
doodlebugs, 13
Farkas, Eugene, 10
Ferguson tractors,
   "Black Tractor," 34
   Ferguson-Brown Model A, 34
   TE-20, 51, 55
   TO-30, 55
   TO-35, 55
Ferguson, Harry, 7, 31, 34, 51, 55
Ford automobiles,
   Model A, 7, 17
   Model B, 10
   Model T, 7, 9, 13, 15, 17
Ford Motor Company, 51
Ford tractor ("Minneapolis" Ford), 20
Ford Tractor Division, 83
Ford Tractor Operations, 83
Ford tractor prototypes,
   Automobile Plow, 9, 10
   Ford-Farkas Prototype, 10
   V-8 prototype (1937), 33
   V-8 prototype (1950s), 65
Fordson tractors, 4, 10, 19–29, 73–81
   Dexta, 73
   E1A New Major, 73
   E27N Major, 73, 79
   Major, 73, 79–81, 84
   Model F, 19, 23, 24, 27
   Model N, 73, 75, 77
   Power Major, 73
   Super Dexta, 73
   Super Major, 73
Ford-Ferguson tractors, 31–49

Model 2N, 31, 41, 47–48, 51, 52, 55, 56, 59
Model 9N, 7, 31, 35, 36, 39–43, 47, 51, 52, 55, 56, 59
Model B-NO-40 Moto-Tug, 48
Model 2NAN, 48
Model 9NAN, 48
Ford tractors,
   Model 8N, 4, 7, 9, 51–61, 63, 65, 68, 71, 95
   Model 2000, 83
   Model 4000, 90, 92
   Model 5000, 83
   Model 600, 83, 84–85, 86
   Model 6000, 83, 90–91
   Model 601, 86–87
   Model 700, 83, 84
   Model 800, 83, 84
   Model 900, 83, 84
   Model 961, 88
   Model 971, 88
   Model NAA Golden Jubilee, 83, 84
Ford-Funk tractors, 63, 67–71
   Model 8N V-8, 63, 67–71, 95
   Model 8N Six-Cylinder, 71
Ford, Henry, 7, 9–10, 19–20, 23, 31, 34, 35, 51, 73
Ford, Henry II, 51
Funk Aircraft Company, 63
Funk, Howard, 63
Funk, Joe, 63
Hadfield-Penfield Steel Co., 27
Henry Ford & Son, 20
Nebraska Tractor Test Law, 20
Pullford, 15
Smith Form-A-Tractor, 14–15
Staude Mak-A-Tractor, 13, 16–17

1950s Ford-Funk Model 8N V-8
Ford V-8 power in the small 8N chassis created a lightweight tractor with all the brawn a farmer could ever wish for. The sole concern was at times having too much muscle under the hood and spinning the rear wheels to sink the tractor into a field's soil. Owner: Palmer Fossum. (Photograph by Chester Peterson Jr.)